Jennifer Munro

Sexual dimorphism & habitat of the scorpion Urodacus elongatus

AF153248

Jennifer Munro

Sexual dimorphism & habitat of the scorpion Urodacus elongatus

Endemic species of the Flinders Ranges, South Australia

LAP LAMBERT Academic Publishing

Impressum / Imprint

Bibliografische Information der Deutschen Nationalbibliothek: Die Deutsche Nationalbibliothek verzeichnet diese Publikation in der Deutschen Nationalbibliografie; detaillierte bibliografische Daten sind im Internet über http://dnb.d-nb.de abrufbar.

Alle in diesem Buch genannten Marken und Produktnamen unterliegen warenzeichen-, marken- oder patentrechtlichem Schutz bzw. sind Warenzeichen oder eingetragene Warenzeichen der jeweiligen Inhaber. Die Wiedergabe von Marken, Produktnamen, Gebrauchsnamen, Handelsnamen, Warenbezeichnungen u.s.w. in diesem Werk berechtigt auch ohne besondere Kennzeichnung nicht zu der Annahme, dass solche Namen im Sinne der Warenzeichen- und Markenschutzgesetzgebung als frei zu betrachten wären und daher von jedermann benutzt werden dürften.

Bibliographic information published by the Deutsche Nationalbibliothek: The Deutsche Nationalbibliothek lists this publication in the Deutsche Nationalbibliografie; detailed bibliographic data are available in the Internet at http://dnb.d-nb.de.

Any brand names and product names mentioned in this book are subject to trademark, brand or patent protection and are trademarks or registered trademarks of their respective holders. The use of brand names, product names, common names, trade names, product descriptions etc. even without a particular marking in this work is in no way to be construed to mean that such names may be regarded as unrestricted in respect of trademark and brand protection legislation and could thus be used by anyone.

Coverbild / Cover image: www.ingimage.com

Verlag / Publisher:
LAP LAMBERT Academic Publishing
ist ein Imprint der / is a trademark of
OmniScriptum GmbH & Co. KG
Heinrich-Böcking-Str. 6-8, 66121 Saarbrücken, Deutschland / Germany
Email: info@lap-publishing.com

Herstellung: siehe letzte Seite /
Printed at: see last page
ISBN: 978-3-659-49965-4

Table of Contents

Sexual Dimorphism and Demography of the Scorpion *Urodacus elongatus* (Scorpionidae)

INTRODUCTION

Scorpions are an ancient group of arachnids historically most represented in folklore and mythology. The combination of four pairs of hairy legs, large pedipalps, and "tail" (metasoma) with a venomous telson has caused them to be feared and reviled. While a great deal is known about scorpion venom, relatively little is known about their life history and demography (Polis 1990). Previous ecological work on scorpions has also been largely restricted to northern hemisphere species (e.g. Benton 1991; Francke 1976; 1979; 1984; Francke and Jones 1982; Lourenco *et al.* 2003; Ozkan *et al.* 2006). Scorpions develop through a series of instars defined by changes in size and shape with moulting of their exoskeleton. The first instar lasts for a few weeks, during which they stay with the mother. Following the first moult, second instar scorpions become independent and disperse. The number of moults is variable among species, e.g. four in *Centruroides arctimanus* (de Armas and Contreras 1981), six in *Opisthacanthus cayaporum* (Lourenco 1985), and eight in *Diplocentrus trinitarius* (de Armas 1982). Time to maturity is extremely variable both within and among species, e.g. 8-10 months for *Centruroides gracilis* (de Armas and Contreras 1981), 19-24 months for *Paruroctonus mesaensis* (Polis and Farley 1979), and 39-83 months for *Pandinus gambiensis* (Vachon et al. 1970).

There are three primary methods of identifying instars for a given species: theoretical; direct; and indirect (Francke and Sissom 1984). The theoretical method is based on a progression law for the regular geometrical incremental size increases of some arthropods (Przibram and Megusar 1912). Thus, from known mass or length of structures on newborns and adults, the number of intervening moults can be calculated. The direct method involves raising captive bred young to maturity and recording the timing of moults and characteristics of each instar. The indirect method involves sampling wild populations and estimating instar numbers and characteristics from clusters of plotted measurements.

1

Morphometric analysis of scorpion instars has been performed on a range of northern hemisphere species (Benton 1991; Francke 1976; Francke and Jones 1982). In the southern hemisphere the scorpion subfamily Urodacinae (Scorpionidae) are endemic to Australia, species distribution ranges from widespread (e.g. *Urodacus yaschenkoi* and *U. manicatus*) to localised (e.g. *U. koolanensis,* and *U. spinatus*) (Koch 1977). *Urodacus* scorpions are under-represented in the literature, and most of the available information is decades old (Shorthouse and Marples 1982; Smith 1966; Woodman 2008). This present study focuses on the Flinders Ranges scorpion (*Urodacus elongatus*). *U. elongatus* is endemic to the region and has patchy distribution associated with gully areas, and constructs short burrows and scrapes underneath rocks (Koch 1977). These scorpions display an extreme level of sexual dimorphism for scorpions (Koch 1977). However the development and particulars of this sexual dimorphism have not been described previously.

Sexual dimorphism can occur in body characteristics such as size, shape and colour (Colonello *et al.* 2007; Dietz *et al.* 2006; Kwiatkowski and Sullivan 2002). In scorpions, specific characteristics include body size, tail length, colour and pectine (sensory structure unique to scorpions) size (Carthy 1968; Polis 1990). Evolution of sexual dimorphism can be driven by mechanisms such as sexual selection and niche segregation. Sexual selection favours larger size in secondary sex traits that increase fitness, e.g. weapons, and predicts lower survivorship among the most divergent age/sex class (Andersson 1994; Baird *et al.* 1997; Moller 1995). In contrast, niche segregation predicts ecological differences in the most divergent age/sex class (Calenge and Basille 2008; Hedrick and Temeles 1989; Pearson *et al.* 2002). The mechanisms driving sexual dimorphism in *U. elongatus* have not been previously examined.

The aims of this paper are to 1) determine how many instars occur in *Urodacus elongatus,* 2) describe the extent and development of the sexual dimorphism, and 3) find out whether/when sexes differ in movement and survivorship.

METHODS

Study Site

This study was done in the southern Flinders Ranges at the Mount Remarkable National Park (MRNP) (32°46'38.20 S 138°04'30.00 E) and Telowie Gorge Conservation Park (TGCP) (32°58'58.00 S 138°05'35.40 E). The MRNP ranges from the western coastal plain near Mambray Creek to the summit of Mount Remarkable in the east, incorporating the Alligator and Mambray Creek catchments. The TGCP is a smaller area approximately 30km south of MRNP and incorporates the Telowie Creek catchment. Vegetation present at both sites is primarily mallee woodland, with some bare scree slopes (D.E.H. 2001; Davies *et al.* 1996). The predominant flora includes *Eucalyptus*, *Callitris glaucophylla* and *Acacia*, native and introduced grasses, and low-growing herbaceous plants (D.E.H. 2001; Davies *et al.* 1996). Rocks are ancient quartzite and sandstone. Soils are generally alkaline, well-drained, reddish, dense loams, shallow on slopes and deeper in gullies (D.E.H. 2001; Davies *et al.* 1996). The regional climate is Mediterranean, with long, hot dry summers (average maximum temperature 30°C), and cool, wet winters (average maximum temperature 20°C). The majority of rainfall occurs from April to October, with variable spatial distribution due to local topography. Regional annual rainfall averages range from 347mm at Mambray Creek in the west to 589mm at Melrose, in the east of MRNP (D.E.H. 2001).

Defining Instars

As there was not enough time to raise scorpions through their entire lifecycle (direct method), I used the indirect method (Francke and Sissom 1984) to identify instars for *U. elongatus*. I used wild caught scorpions from MRNP and TGCP, and wet-preserved specimens from the South Australian Museum reference collection. Anatomical measurements (adapted from Koch 1977, Fig. 1) were the basis from which instars were defined. Claw base (CIBL) and tip (CITL) length, claw width (CIW), medial carapace length (MCaL) and width (MCaW), metasomal fourth segment length (TL), pectine length (PL) and teeth number (PTNo) were measured for all scorpions. The fifth metasomal segment is most sexually dimorphic however standard methods incorporating the fourth segment were used in this study (Koch 1977, Fig. 1).

3

Measurements of wild scorpions were made using digital display callipers (±0.05mm). Museum specimen measurements were made using a dissecting microscope (10 x magnifications) with a graduated scale (0.1mm) eyepiece. Sex was determined for all scorpions using a dissecting microscope (10x magnification) to inspect the genital orifice for presence/absence of male genital papilla (Polis 1990).

Population sampling

A focused population study of *U. elongatus* was undertaken in a 45x45m area of the Mambray Creek floodplain near the Mambray Creek entrance to MRNP (32°50 .527' S 138° 01.555' E). The site included low-lying, dry flood banks along the northern and southern boundaries that represent suitable habitat for *U. elongatus* (J. Munro, unpublished data, 2008).

The population was sampled over two nights during ten field trips from March to August 2008. Searches commenced within an hour of sunset and a portable UV light was used to systematically scan the area for 2-2.5 hours. Any active scorpions were included, captured, measured and marked individually. Active scorpions were defined as either: 1) found in open away from burrow; or 2) at entrance to burrow. Scorpions were placed individually into zip-lock bags using 30cm forceps, ready for processing and return to the place of capture the following day. Scorpions released in the open were watched until they moved to shelter beneath the nearest adjacent rock, to prevent them being predated. Scorpions collected during the two nightly searches for each field trip were pooled as a single sample.

Scorpions were marked with paint dots representing units, tens, and hundreds, so that they had a unique identifying number (Fig. 2). Paint used was water-based, non-toxic Micador Acrylic Paint. Marks persisted throughout the study period (J. Munro, pers. obs., 2008).

Records for individual scorpions were used to construct a calendar of catches (Petrusewicz and Andrzejewski 1962, Appendix 2). In order to maximise information gained, the following definitions and simplifying assumptions were made:

1. Autumn-spring timing of the study was assumed to rule out loss of paint dots due to moulting that is reported to occur during summer (M. Newton, pers. obs., 2007).

4

2. The smallest size class collected possessing a hardened exoskeleton, was assumed to represent the first independent (second) instar.

3. All unmarked scorpions found were assumed to be previously uncaptured, excluding any found with a recently moulted marked exoskeleton.

4. Marked individuals found in the same burrow but with breaks in the capture record were assumed to have been living in that burrow during the intervening period.

5. Marked individuals found in new burrow locations were assumed to have moved to that burrow only within the period immediately preceding that census.

Scorpion capture locations were plotted on a map of the site. To identify movement patterns for individuals, the distance moved between capture points was calculated. Intersexual differences in captures and movement of instars were collated for statistical analysis. Raw data were entered using Microsoft Excel 2007 software.

Statistical Analysis

The number of instars was estimated from scatter plots of raw data for MCal against total chela length (CIBL+CITL). Discriminant function analysis (Norusis 1990) was performed separately for males, females, and for the two sexes combined, using the raw data obtained for CIBL, CITL, CIW, MCaL, MCaW, TL, and PL (Tables 2-4). A priori groupings of second through to sixth instars were used in the analysis. Canonical discriminant analysis was used to create a robust method for determining instar distinction for individual *U. elongatus,* and explore development and nature of sexual dimorphism.

In order to establish whether the instars of each sex differed in frequency of occurrence within the population a frequency tower was constructed from the total numbers of males and females caught of each instar (Fig. 7). Chi-square tests were then used to determine significance of differences in capture frequencies between the sexes for each instar. For sixth instar scorpions Fisher's exact probability tests were used to test the hypothesis that the sexes were equally likely to be recaptured. I used Mann-Whitney U test to test the hypothesis that mature individuals (instar = 6) of both sexes did not differ in the maximum distances they travelled during the study.

All statistical analysis was performed using SPSS 15.0 for Windows.

RESULTS

Instars of Urodacus elongatus

The scatter plots of MCal against CIBL+CITL indicated five distinct size classes (instars) for both males (n=123) (Fig. 3) and females (n=171) (Fig. 4). The raw morphometric data for the variables CIBL, CITL, CIBL+CITL, CIW, MCaL, MCaW, TL, and PL were used to calculate mean(±S.D.) values over the range for all five instars of each sex (Table 1).

Identification of Male instars

Discriminant function analysis with the five male instars as a priori groups resulted in two significantly discriminant functions that incorporated seven (CIBL, CITL, CIW, MCaL, MCaW, TL, and PL) of the eight variables entered into the analysis (Table 2). All measurements contributed significantly to decrease Wilks' Lambda except CIBL+CITL, so it was not included in the analysis. These functions accounted for 98.1% and 1.6% of the variance, respectively (P<0.001) (Table 2). Within-group correlations of the seven measurements were generally high and positive (Table 2). The analysis resulted in correct identification of 98.4% of male scorpions within the five predicted instars. A priori groups were plotted using the results of canonical discriminant analysis (Fig. 5). Five distinct groups were clearly defined coinciding with the axis of function one, however no clear distribution pattern was evident associated with function two.

Identification of Female instars

Discriminant function analysis with the five female instars as a priori groups resulted in one significantly contributing function that incorporated seven (CIBL, CITL, CIW, MCaL, MCaW, TL, and PL) of the eight variables entered into the analysis (Table 3). All measurements contributed significantly to decrease Wilks' Lambda except CIBL+CITL, so it was not included in the analysis. The function accounted for 99.4% of the variance (P<0.001) (Table 3). Within-group correlations of the seven measurements were generally high and positive (Table 3). The analysis resulted in correct identification of 97.6% of female scorpions within the five predicted instars.

6

Development of Sexual Dimorphism

For the combined ten instars as a priori groups for the two sexes, discriminant analysis resulted in three significantly contributing functions that incorporated the same seven variables as individual sex analyses (Table 4). These functions accounted for 75.8%, 23.3%, and 0.7% of the variance respectively ($P<0.001$) (Table 4). The analysis resulted in correct identification of 88% of scorpions of both sexes within the ten predicted instars. Therefore, separate sex discriminant analyses provided more accurate (98.4% and 97.6%) assignment of scorpions within a priori groups than the pooled sex discriminant analysis. A priori groups of both sexes were plotted using the results of canonical discriminant analysis (Fig. 6). Divergence between the sexes was present in increasing magnitude from the fourth instars to the sixth instars.

Although sexual dimorphism of general body shape does not occur until the sixth instar, sexual differences in the pectines were consistently evident in all instars. Mean ± S.D. pectine lengths remained larger in males than females (Table 1), as did the number of pectine teeth in males (mean = 20) compared with females (mean = 13).

Population Structure and Movement

Significant intersexual differences in capture abundances of instars were detected ($X^2_4=9.87$, $P<0.05$). The population frequency tower demonstrated approximately symmetrical captures between the sexes in the first four instars (P values>0.05). Therefore, significance was due solely to intersexual differences in sixth instars ($X^2_1=4.77$, $P<0.05$), where approximately three times more females than males were captured (Fig. 7).

Differences in recaptures (7/14 males and 31/39 females) were suggestive of a trend more strongly influencing sixth instar males. However, these differences were not significant ($X^2_1=0.41$, $P>0.5$). Maximum distance travelled between captures for sixth instars did not differ between males (0-20m, $n=14$) and females (0-34m, $n=39$). Mean ± S.D. values for females (3.68±7.15m) and males (2.62±5.96m) were found to be non-significant ($z=-1.082$, $P=0.279$) when a Mann-Whitney U test was applied. This suggests the trend could not be explained by differences in dispersal.

DISCUSSION

Life History

Five distinct size classes were identified for each sex via indirect morphometric analysis (Figs 1, 3 and 4). I interpret these to represent instars two to six for *U. elongatus*. Indirect methods of analysis have previously been used in other scorpion species to identify instars (Benton 1991; Francke 1979; 1984) including other *Urodacus* scorpions (Shorthouse and Marples 1982; Smith 1966). Comparable morphometric techniques have also been used to describe the life history of other taxa, such as red-spot prawn (Tzeng *et al.* 2001); dolphins (Sanvicente-Anorve *et al.* 2004); and mole-rats (Hart *et al.* 2007). Identifying size classes for a species provides an understanding of growth and maturation patterns. In conjunction, establishing a set of measurable criteria via discriminant functions for each size class enables future individuals to be assigned to a specific class. For future study of *U. elongatus* confidence in assigning individuals to instars, will provide more detailed information about population structure and dynamics.

Scorpion populations are comprised of individuals from a range of instars, that each grow during a moult (Polis 1990). Lifetime number of moults is variable among scorpions, ranging from four to nine (Polis 1990). Previous studies of *Urodacus* species have reported six instars (five moults) in *U. yaschenkoi* (Shorthouse and Marples 1982) and *U. manicatus* (=*U. abruptus*) (Smith 1966). The current study has confirmed the presence of six instars in *U. elongatus* also (Figs. 3 and 4). While these results seem clear, confirmation using the direct method of raising individuals through the entire lifecycle is desirable (Francke and Sissom 1984).

The primary significant discriminant function in both sexes showed high, positive within-group correlations, suggesting that this function relates to increase in size (Tables 2 and 3). Development of the sexual dimorphism in *U. elongatus* corresponds to attainment of the final sixth instar. *U. elongatus* sixth instars show extraordinary sexual dimorphism for scorpions (Koch 1977). Sixth instar males in particular, deviate from the growth pattern observed in immature instars (Figs. 5 and 6). Sexual maturity in scorpions occurs after the final or penultimate moult, however age at maturity is highly variable (Koch 1977; Polis 1990). Maturation occurs after five-six moults at eight-ten months in *C. gracilis* (Francke and Jones 1982), after four-five moults at 25-27 months in *Tityus fasciolatus* (Lourenco 1978) and after eight-

8

nine moults at 47 months in *D. trinitarius* (de Armas 1982). In *Urodacus* scorpions, *U. yaschenkoi* mature after five moults at 54 months, compared with five moults at 18-25 months in *U. manicatus* (Shorthouse and Marples 1982; Smith 1966). This variation suggests caution in assigning age at maturity for *U. elongatus*. Future research on *U. elongatus* should use direct methods to investigate instar timing, and how this varies with factors such as seasonal temperature.

Number of moults to maturity may also vary intersexually in scorpions, e.g. *Orthochirus innesi* (Shulov and Amitai 1960) males (4) and females (5), and *D. trinitarius* males (8) and females (9) (de Armas 1982). Maturity of both sexes in *Urodacus* species occurs at the sixth instar (Shorthouse and Marples 1982; Smith 1966), and the presence of extreme sexual dimorphism in *U. elongatus* in only the sixth instars suggests they also conform to this pattern. Absence of post-maturation moults in scorpions renders the adult instar class, a cumulative group (Polis 1990; Shorthouse and Marples 1982). The degree of accumulation comparative to numbers of immature individuals is indicative of species longevity. High frequencies of sixth instar female *U. elongatus* captured probably represent an accumulation of long-lived adults (Fig. 7). Dissections of mature female reproductive systems in *Urodacus* scorpions show embryonic evidence of at least three episodes of reproductive output (Smith 1990; Warburg and Rosenberg 1994). Replication of this technique in future study of *U. elongatus* would assist in confirming longevity and reproductive output.

Sex Dimorphism

Combined sex canonical discriminant analysis (Fig. 6) showed that sexual dimorphism increased through subsequent instars, becoming most notable in the final instar. *U. elongatus* displayed consistent sexual differences in pectine length and pectine teeth number. In all instars these characters appear to be primary sexual traits, i.e. those essential for reproductive function and present prior to maturity (Darwin 1871). Contrasting with this, sexual dimorphism in body shape developed at maturity. Sexual dimorphism in scorpions is usually due to fecundity selection for larger female body size linked with increased offspring number in larger females (Brown 2004). This offers an explanation for the female size bias observed in mature *U. elongatus* (Table 1). Female-biased sexual dimorphism is more common in ectotherms, e.g. reptiles (Baird *et al.* 2003; Butler and Losos 2002); insects (House and Simmons 2003); and fish (Reimchen and Nosil 2004). In contrast

9

male-biased sexual dimorphism is widely reported in endotherms, e.g. birds (Badyaev and Hill 2000a; Dunn *et al.* 2001); and mammals (Garel *et al.* 2006; Mazak 2004). Male-biased sexual dimorphism is commonly due to sexual selection for traits, e.g. weapons, which can increase offspring survival and thereby parental fitness (Andersson 1994).

In contrast to the patterns of female-bias in sexual size dimorphism in other ectotherms, mature male *U. elongatus* scorpions display the most extreme sexual dimorphism in shape. The second significant discriminant function in the males did not show a clear pattern for instars, but probably relates to changes in shape due to maturation (Table 2, Fig. 5). The presence of longer metasomal segments in mature males is likely the primary character responsible (Koch 1977). This distinctive trait could relate to increased competitiveness of longer-tailed males in intrasexual territorial conflicts. This explanation would support evolution of the male-biased sexual dimorphism due to sexual selection (Badyaev *et al.* 2000b; Baird *et al.* 2003; Dunn *et al.* 2001). Alternatively, longer male tails could relate to protection from sexual cannibalism, a common phenomenon in scorpions (Polis 1990). Natural selection favours optimal morphology, supporting this explanation (Butler and Losos 2002; House and Simmons 2003). The degree to which a sex dimorphism becomes established in a species will reflect trade-offs between sexual and natural selection, in combination with intersexual differences in survival and dispersal (Bonduriansky 2007; Ritchie *et al.* 2007).

Population Dynamics

Significant intersexual differences in catchability of the different instars were found in the sixth. Low and almost symmetrical captures for second to fifth instars indicate that birth sex ratios are likely to approximate 1:1, and that survival and dispersal to fifth instar is likely to be similar intersexually (Fig. 7). This assumes equal numbers of immigrants/emigrants and catchability for both sexes. In comparison, in the sixth instars almost three times more females than males were captured (Fig. 7). The same female-biased 3:1 adult sex ratio was also reported in adult *U. manicatus* (Smith 1966), and 2:1 in *U. yaschenkoi* (Shorthouse and Marples 1982). These ratios indicate that intersexual survival and/or dispersal is likely different in adults. Recaptures and dispersal of sixth instars of both sexes did not differ significantly in *U. elongatus*, indicating both sexes were similarly sampled by recaptures, and were similarly dispersing. Comparatively lower frequency of sixth instar males

10

suggests that mature males may either, remain more consistently within the burrow, utilise larger home ranges or ranges further from the natal population, or are experiencing higher mortality. Previous literature on scorpions confirms that behaviours such as burrowing and nocturnal activity utilised by *U. elongatus* provide protection against risks of desiccation and predation, supporting the first possibility (Bradley 1982; Bradley 1988; Cloudsley-Thompson 1978; Crawford and Krehoff 1975; Polis 1980). Alternatively, larger amounts of exposed surface activity during establishment and maintenance of larger home ranges would render males more susceptible to higher mortality by natural selection, supporting the latter two possibilities.

Simultaneously, more consistent residency of mature females suggests that males are the sex actively searching for mates. This male-biased pattern of mate searching has also been reported in other scorpions (Benton 1992; Polis and Farley 1980). Tendency for mature males to be the more nomadic sex resulting in smaller numbers of residential individuals has also been reported in birds (Wheelwright *et al.* 1994) and mammals (Boydston *et al.* 2005). In raptors in particular, males also face strong sexual selection for smaller size as the more active provisioning sex, to increase foraging efficiency (Andersson and Norberg 1981). Sexual size dimorphism in *U. elongatus* may also relate to role or resource partitioning, supporting intersexual niche segregation. Dispersal distance in males and intersexual genetic differences within populations, have been positively linked to sexual dimorphism in fish (Ritchie *et al.* 2007). Future study of *U. elongatus* should incorporate a component of population intersexual genetic diversity to assist in quantifying differences in dispersal.

Conclusions

This study has shown that there are six instars for male and female *U. elongatus,* and provides a discriminant function analysis that allows determination of instars based on morphometrics. The genital papilla and pectine morphology enable the sexes to be differentiated in second to sixth instars. An extreme sexual dimorphism in shape occurs in this species, with males being longer than females. This sexual dimorphism develops slowly through the second to fifth instars, and is only fully expressed in sixth instars. Predominance of the dimorphic traits in the males is suggestive of sexual selection for these traits. Sex ratios and intersexual activity of instars was comparable in the second to fifth, and were significantly different in the sixth.

Lower frequency of mature males than females, combined with no significant difference in dispersal, suggests males are more at risk of mortality by natural selection processes. The mechanisms giving rise to the extreme sexual dimorphism in *U. elongatus* are unknown, and deserve further attention.

REFERENCES

Andersson M. (1994) *Sexual Selection*. Princeton University Press, Princeton.

Andersson M. & Norberg R. A. (1981) Evolution of reversed sexual size dimorphism and role partitioning among predatory birds, with a size scaling of flight performance. *Biological Journal of the Linnean Society* **15**, 105-30.

Badyaev A. V. & Hill G. E. (2000) The evolution of sexual dimorphism in the house finch. I. Population divergence in morphological covariance structure. *Evolution* **54**, 1784-94.

Badyaev A. V., Hill G. E., Stoehr A. M., Nolan P. M. & McGraw K. J. (2000) The evolution of sexual size dimorphism in the house finch. II. Population divergence in relation to local selection. *Evolution* **54**, 2134-44.

Baird T. A., Fox S. F. & McCoy J. K. (1997) Population differences in the roles of size and coloration in intra- and intersexual selection in the collared lizard, *Crotaphytus collaris*: influence of habitat and social organization. *Behavioral Ecology* **8**, 506-17.

Baird T. A., Vitt L. J., Baird T. D., Cooper W. E., Caldwell J. P. & Perez-Mellado V. (2003) Social behavior and sexual dimorphism in the Bonaire whiptail, *Cnemidophorus murinus* (Squamata : Teiidae): the role of sexual selection. *Canadian Journal of Zoology-Revue Canadienne De Zoologie* **81**, 1781-90.

Benton T. G. (1991) The life history of *Euscorpius flavicaudis* (Scorpiones, Chactidae). *Journal of Arachnology* **19**, 105-10.

Benton T. G. (1992) The ecology of the scorpion *Euscorpius flavicaudis* in England. *Journal of Zoology, London* **226**, 351-68.

Bonduriansky R. (2007) Sexual selection and allometry: A critical reappraisal of the evidence and ideas. *Evolution* **61**, 838-49.

Boydston E. E., Kapheim K. M., Van Horn R. C., Smale L. & Holekamp K. E. (2005) Sexually dimorphic patterns of space use throughout ontogeny in the spotted hyena (*Crocuta crocuta*). *Journal of Zoology* **267**, 271-81.

Bradley R. (1982) Digestion time and re-emergence in the desert grassland scorpion *Paruroctonus utahensis* (Williams) (Scorpionida, Vaejovidae). *Oecologia* **55**, 316-8.

Bradley R. A. (1988) The influence of weather and biotic factors on the behaviour of the scorpion (*Paruroctonus utahensis*). *Journal of Animal Ecology* **57**, 533-51.

Brown C. A. (2004) Life histories of four species of scorpion in three families (Buthidae, Diplocentridae, Vaejovidae) from Arizona and New Mexico. *Journal of Arachnology* **32**, 193-207.

Butler M. A. & Losos J. B. (2002) Multivariate sexual dimorphism, sexual selection, and adaptation in Greater Antillean Anolis lizards. *Ecological Monographs* **72**, 541-59.

Calenge C. & Basille M. (2008) A general framework for the statistical exploration of the ecological niche. *Journal of Theoretical Biology* **252**, 674-85.

Carthy J. D. (1968) The pectines of scorpions. *Symposium of the zoological society of London* **23**, 251-61.

Cloudsley-Thompson J. L. (1978) Biological clocks in Arachnida. *Bulletin of the British Arachnological Society* **4**, 184-91.

Colonello J. H., Lucifora L. O. & Massa A. M. (2007) Reproduction of the angular angel shark (*Squatina guggenheim*): geographic differences, reproductive cycle, and sexual dimorphism. *Ices Journal of Marine Science* **64**, 131-40.

Crawford C. S. & Krehoff R. C. (1975) Diel activity in sympatric populations of the scorpions *Centruroides sculpturatus* (Buthidae) and *Diplocentrus spitzeri* (Diplocentridae). *Journal of Arachnology* **2**, 195-204.

D.E.H. (2001) *Mount Remarkable National Park Management Plan - Draft*. Department for Environment and Heritage, Adelaide, South Australia.

Darwin C. (1871) *The Descent of Man and Selection in Relation to Sex*. John Murray.

Davies M., Twidale C. R. & Tyler M. J. (1996) *Natural History of the Flinders Ranges*. Royal Society of South Australia Inc., Adelaide.

de Armas L. F. (1982) Desarrollo postembrionario de *Didymocentrus trinitarius* (Franganillo) (Scorpiones: Diplocentridae). *Academia de ciencias de Cuba, Miscelanea zoologica* **16**, 3-4.

de Armas L. F. & Contreras H. (1981) Gestacion y desarrollo postembrionario en algunos *Centruroides* (Scorpionida: Buthidae) de Cuba. *Poeyana* **217**, 1-10.

Dietz C., Dietz I. & Siemers B. M. (2006) Wing measurement variations in the five European horseshoe bat species (Chiroptera: Rhinolophidae). *Journal of Mammalogy* **87**, 1241-51.

Dunn P. O., Whittingham L. A. & Pitcher T. E. (2001) Mating systems, sperm competition, and the evolution of sexual dimorphism in birds. *Evolution* **55**, 161-75.

Francke O. F. (1976) Observations on the life history of *Uroctonus mordax* Thorell (Scorpionida, Vaejovidae). *Bulletin of the British Arachnological Society* **3**, 254-60.

Francke O. F. (1979) Observations on the reproductive biology and life history of *Megacormus gertschi* Diaz (Scorpiones: Chactidae; Megacorminae). *Journal of Arachnology* **7**, 223-30.

Francke O. F. (1984) The life history of *Diplocentrus bigbendensis* Stahnke (Scorpiones, Diplocentridae). *The Southwestern Naturalist* **29**, 387-93.

Francke O. F. & Jones S. K. (1982) The life history of *Centruroides gracilis* (Scorpiones, Buthidae). *Journal of Arachnology* **10**, 223-39.

Francke O. F. & Sissom W. D. (1984) Comparative review of the methods used to determine the number of molts to maturity in scorpions (Arachnida), with analysis of the post-birth development of *Vaejovis coahuilae* Williams (Vaejovidae). *Journal of Arachnology* **12**, 1-20.

Garel M., Solberg E. J., Saether B. E., Herfindal I. & Hogda K. A. (2006) The length of growing season and adult sex ratio affect sexual size dimorphism in moose. *Ecology* **87**, 745-58.

Hart L., Chimimba C. T., Jarvis J. U. M., O'Riain J. & Bennett N. C. (2007) Craniometric sexual dimorphism and age variation in the South African Cape dune mole-rat (*Bathyergus suillus*). *Journal of Mammalogy* **88**, 657-66.

Hedrick A. V. & Temeles E. J. (1989) The evolution of sexual dimorphism in animals: Hypotheses and tests. *Trends in Ecology & Evolution* **4**, 136-8.

House C. M. & Simmons L. W. (2003) Genital morphology and fertilization success in the dung beetle *Onthophagus taurus*: an example of sexually selected male genitalia. *Proceedings of the Royal Society of London Series B-Biological Sciences* **270**, 447-55.

Koch L. (1977) The taxonomy, geographic distribution and evolutionary radiation of Australo-Papuan scorpions. *Records of the Western Australian Museum* **5**, 83-367.

Kwiatkowski M. A. & Sullivan B. K. (2002) Geographic variation in sexual selection among populations of an iguanid lizard, *Sauromalus obesus* (=*ater*). *Evolution* **56**, 2039-51.

Lourenco W. R. (1978) Etude sur les scorpions appartenants au "complexe" *Tityus trivittatus* Kraepelin, 1898, et, en particulier, de la sous-espece *Tityus trivittatus fasciolatus* Pessoa, 1935 (Buthidae). p. 128. L'Universite Pierre et Marie Curie, Paris.

Lourenco W. R. (1985) Essai d'interpretation de la distribution du genre *Opisthacanthus* (Arachnida, Scorpiones, Ischnuridae) dans les regions Neotropicale et Afrotropicale. Etude taxinomique, biogeographique, evolutive et ecologique. p. 287. L'Universite Pierre et Marie Curie, Paris.

Lourenco W. R., Andrzejewski V. & Cloudsley-Thompson J. L. (2003) The life history of *Chactas reticulatus* (Kraepelin, 1912) (Scorpiones, Chactidae), with a comparative analysis of the reproductive traits of three scorpion lineages in relation to habitat. *Zoologischer Anzeiger* **242**, 63-74.

Mazak J. H. (2004) On the sexual dimorphism in the skull of the tiger (*Panthers tigris*). *Mammalian Biology* **69**, 392-400.

Moller A. P. (1995) Sexual selection in the barn swallow (*Hirundo rustica*). 5. Geographic variation in ornament size. *Journal of Evolutionary Biology* **8**, 3-19.

Norusis M. J. (1990) *SPSS Advanced Statistics User's Guide*. SPSS Inc., Chicago.

Ozkan O., Adiguzel S. & Kar S. (2006) Parametric values of *Androctonus crassicauda* (Oliver, 1807) (scorpiones: buthidae) from Turkey. *Journal of Venomous Animals and Toxins Including Tropical Diseases* **12**, 549-59.

Pearson D., Shine R. & How R. (2002) Sex-specific niche partitioning and sexual size dimorphism in Australian pythons (*Morelia spilota imbricata*). *Biological Journal of the Linnean Society* **77**, 113-25.

Petrusewicz K. & Andrzejewski R. (1962) Natural history of a free-living population of house mice (*Mus musculus L.*) with particular reference to grouping within the population. *Ekologia Polska Seria A* **10**, 85-122.

Polis G. A. (1980) Seasonal Patterns and Age-Specific Variation in the Surface Activity of a Population of Desert Scorpions in Relation to Environmental Factors. *The Journal of Animal Ecology* **49**, 1-18.

Polis G. A. (1990) *The Biology of Scorpions*. Stanford University Press, Stanford.

Polis G. A. & Farley R. D. (1979) Characteristics and environmental determinants of natality, growth and maturity in a natural population of the desert scorpion, *Paruroctonus mesaensis* (Scorpionida: Vaejovidae). *Journal of Zoology (London)* **187**, 517-42.

Polis G. A. & Farley R. D. (1980) Population biology of a desert scorpion: Survivorship, microhabitat, and the evolution of life history strategy. *Ecology* **61**, 620-9.

Przibram H. & Megusar F. (1912) Wachstummessungen an *Sphodromantis bioculata* Burm. 1. Lange und Masse. *Arch. EntwMech. Org.* **34**, 680-741.

Reimchen T. E. & Nosil P. (2004) Variable predation regimes predict the evolution of sexual dimorphism in a population of three spine stickleback. *Evolution* **58**, 1274-81.

Ritchie M. G., Hamill R. M., Graves J. A., Magurran A. E., Webb S. A. & Garcia C. M. (2007) Sex and differentiation: population genetic divergence and sexual dimorphism in Mexican goodeid fish. *Journal of Evolutionary Biology* **20**, 2048-55.

Sanvicente-Anorve L., Lopez-Sanchez J. L., Aguayo-Lobo A. & Medrano-Gonzalez L. (2004) Morphometry and sexual dimorphism of the coastal spotted dolphin, *Stenella attenuata graffmani*, from Bahia de Banderas, Mexico. *Acta Zoologica* **85**, 223-32.

Shorthouse D. J. & Marples T. G. (1982) The life stages and population dynamics of an arid zone scorpion (*Urodacus yaschenkoi*) (Birula 1903). *Austral Ecology* **7**, 109-18.

Shulov A. & Amitai P. (1960) Observations sur les scorpions: *Orthochirus innesi* E. Sim., 1910, ssp. *negebensis* nov. *Archives de l'Institut Pasteur d'Algerie* **38**, 117-29.

Smith G. T. (1966) Observations on the life history of the scorpion *Urodacus abruptus* (Scorpionidae), and the analysis of its home sites. *Australian Journal of Zoology* **14**, 383-98.

Smith G. T. (1990) Potential lifetime fecundity and the factors affecting annual fecundity in *Urodacus armatus* (Scorpiones, Scorpionidae). *Journal of Arachnology* **18**, 271-80.

Tzeng T. D., Chiu C. S. & Yeh S. Y. (2001) Morphometric variation in red-spot prawn (*Metapenaeopsis barbata*) in different geographic waters off Taiwan. *Fisheries Research* **53**, 211-7.

Vachon M., Roy R. & Condamin M. (1970) Le developpement postembryonnaire du scorpion *Pandinus gambiensis* Pocock. *Bulletin de l'Institut francais d'Afrique noire* **32**, 412-32.

Warburg M. R. & Rosenberg M. (1994) The female reproductive system of the eastern Australian scorpion. *Tissue and Cell* **26**, 779-83.

Wheelwright N. T., Trussell G., Devine J. P. & Anderson R. (1994) Sexual dimorphism and population sex-ratios in juvenile savannah-sparrows. *Journal of Field Ornithology* **65**, 520-9.

Woodman J. D. (2008) Living in a shallow burrow under a rock: Gas exchange and water loss in an Australian scorpion. *Journal of Thermal Biology* **33**, 280-6.

Table 1. Comparative morphometrics of instars by sex of *Urodacus elongatus* scorpions. Values are given as mean(SD) over range. CIBL=claw base length; CITL=claw tip length; CIBL+CITL= total chela length; CIW=claw width; MCaL=medial carapace length; MCaW=medial carapace width; TL=4th metasomal segment length; PL=pectine length.

Sex & Instar	N	CIBL	CITL	CIBL+ CITL	CIW	MCaL	MCaW	TL	PL
M2	13	2.92(0.19)	3.46(0.30)	6.38(0.32)	1.82(0.26)	3.37(0.32)	3.20(0.22)	2.08(0.28)	1.96(0.33)
		2.6-3.2	3.0-4.2	6.0-7.0	1.6-2.6	2.9-3.9	2.9-3.5	1.7-2.6	1.4-2.7
F2	13	2.79(0.57)	3.49(0.46)	6.27(0.95)	1.73(0.37)	3.30(0.48)	3.16(0.57)	2.13(0.39)	1.78(0.22)
		1.2-3.4	2.4-4.0	3.6-7.4	0.7-2.1	2.1-3.9	1.8-4.2	1.5-2.7	1.4-2.3
M3	18	3.83(0.38)	4.59(0.33)	8.43(0.47)	2.30(0.23)	4.47(0.24)	4.17(0.25)	2.78(0.41)	2.67(0.23)
		3.2-4.3	3.9-5.0	7.4-9.2	2.0-3.0	4.0-5.0	3.9-4.6	2.4-3.8	2.3-3.1
F3	22	4.01(0.40)	4.70(0.46)	8.71(0.62)	2.34(0.23)	4.54(0.27)	4.29(0.34)	2.74(0.35)	2.37(0.39)
		3.0-5.0	3.9-5.4	7.7-10.3	2.0-2.8	3.9-4.9	3.6-5.0	2.3-3.5	1.8-3.3
M4	31	5.31(0.69)	6.12(0.61)	11.43(1.12)	3.11(0.36)	6.01(0.38)	5.55(0.51)	4.29(1.36)	3.83(0.58)
		4.3-6.8	4.8-7.6	9.6-14.2	2.6-4.0	5.1-6.8	4.8-6.8	3.0-8.1	2.7-5.0
F4	28	5.33(0.63)	6.37(0.81)	11.70(1.18)	3.38(0.59)	6.14(0.55)	5.77(0.63)	3.91(0.79)	3.07(0.40)
		3.9-6.9	4.6-7.6	9.5-13.8	2.4-5.0	5.3-7.0	4.8-7.4	2.7-5.5	2.3-3.9
M5	25	7.36(0.66)	8.20(0.70)	15.56(1.15)	4.32(0.43)	7.76(0.46)	7.33(0.48)	7.06(2.04)	5.64(0.63)
		5.7-8.3	6.8-9.9	13.4-18.2	3.7-5.4	7.0-8.5	6.3-8.3	4.4-10.6	4.4-7.4
F5	19	6.98(0.61)	8.30(0.51)	15.29(0.94)	4.67(0.42)	7.91(0.41)	7.35(0.43)	5.33(0.57)	4.14(0.49)
		6.1-8.2	7.4-9.2	13.9-17.4	4.1-5.5	7.0-8.7	6.6-8.3	4.3-6.3	2.9-5.0
M6	36	9.10(0.68)	10.34(0.87)	19.44(1.24)	5.33(0.60)	9.34(0.58)	8.91(0.64)	11.10(1.34)	7.32(0.77)
		7.9-11.1	8.4-12.4	17.6-23.5	4.5-7.3	8.3-11.0	7.6-10.7	8.5-14.4	5.5-8.9
F6	89	9.66(0.90)	11.12(1.03)	20.78(1.58)	6.53(0.76)	10.49(0.78)	9.92(078)	7.12(0.90)	5.46(0.60)
		7.5-11.7	8.7-13.7	17.2-24.8	4.9-9.2	8.5-12.3	7.5-12.0	5.3-9.2	3.4-8.0

Table 2. Unstandardised discriminant function coefficients (and pooled-within-groups correlations with discriminant functions) of seven measurements of five a priori groups of instars of male *Urodacus elongatus* scorpions. MCaL=medial carapace length; MCaW=medial carapace width; CITL=claw tip length; CIBL=claw base length; PL=pectine length; CIW=claw width; TL=4th metasomal segment length.

Variable	Discriminant Function	
	1	2
MCaL	1.428(0.947)	-1.297(-0.110)
MCaW	0.635(0.827)	-0.414(0.091)
CITL	0.428(0.733)	0.064(0.305)
CIBL	0.124(0.704)	-0.054(0.150)
PL	-0.131(0.644)	0.483(0.372)
CIW	-0.194(0.583)	0.121(0.232)
TL	-.007(0.488)	0.720(0.847)
constant	-16.372	4.022
% of variance	98.1	1.6

Table 3. Unstandardised discriminant function coefficients (and pooled-within-groups correlations with discriminant functions) of seven measurements of five a priori groups of instars of female *Urodacus elongatus* scorpions. MCaL=medial carapace length; MCaW=medial carapace width; CITL=claw tip length; CIBL=claw base length; PL=pectine length; CIW=claw width; TL=4th metasomal segment length.

Variable	Discriminant Function
	1
MCaL	0.672(0.934)
MCaW	0.314(0.887)
CITL	0.336(0.759)
CIBL	0.476(0.780)
PL	0.116(0.627)
CIW	-0.189(0.693)
TL	-0.099(0.596)
constant	-13.382
% of variance	99.4

Table 4. Unstandardised discriminant function coefficients (and pooled-within-groups correlations with discriminant functions) of seven measurements of five a priori groups of instars of both sexes of *Urodacus elongatus* scorpions. MCaL=medial carapace length; MCaW=medial carapace width; CITL=claw tip length; CIBL=claw base length; PL=pectine length; CIW=claw width; TL=4th metasomal segment length.

| Variable | Discriminant Function | | |
	1	2	3
MCaL	0.867(0.941)	-0.219(0.065)	-0.542(-0.092)
MCaW	0.369(0.872)	-0.108(0.078)	-0.382(0.083)
CITL	0.411(0.758)	-0.194(0.118)	0.498(0.247)
CIBL	0.354(0.736)	-0.137(0.154)	0.017(0.004)
PL	-0.114(0.550)	1.537(0.644)	-1.514(-0.328)
CIW	0.073(0.685)	-1.235(-0.055)	0.663(0.261)
TL	-0.132(0.445)	0.761(0.610)	0.859(0.610)
constant	-14.122	-0.957	1.607
% of variance	75.8	23.3	0.7

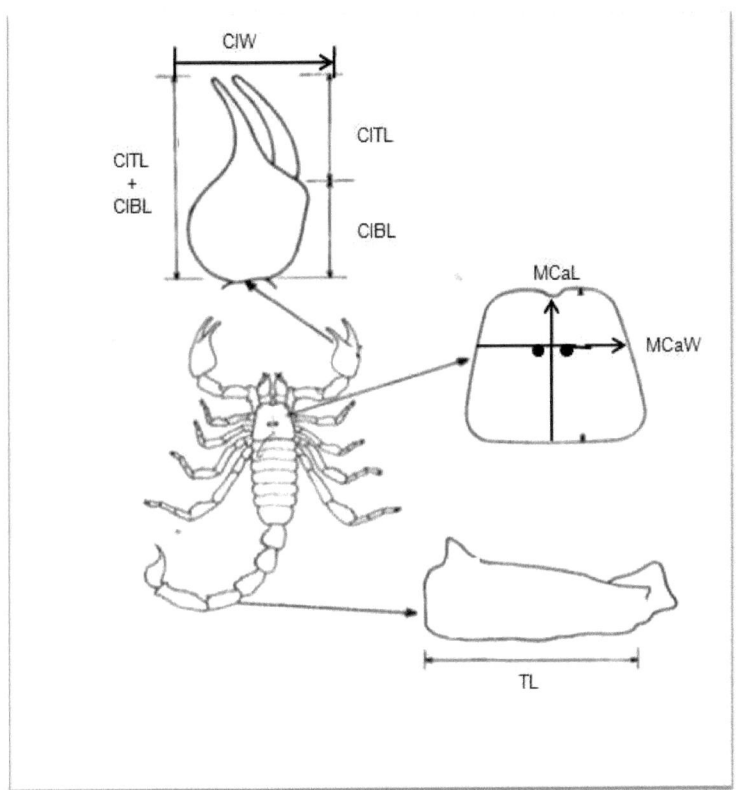

Fig. 1. Morphometrics used to identify age classes and sex dimorphism in *Urodacus elongatus* (adapted from Koch 1977). CITL=claw tip length; CIBL=claw base length; CITL+CIBL=total chela length; CIW=claw width; MCaL=medial carapace length; MCaW=medial carapace width; TL=4th metasomal segment length.

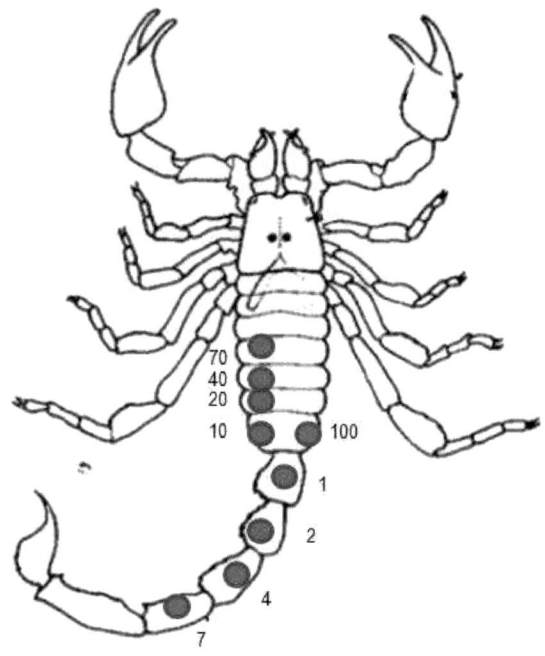

Fig. 2. Numbering system used for individual identification of *Urodacus elongatus*. A combination of units, tens, and hundreds was used to give each individual a unique number.

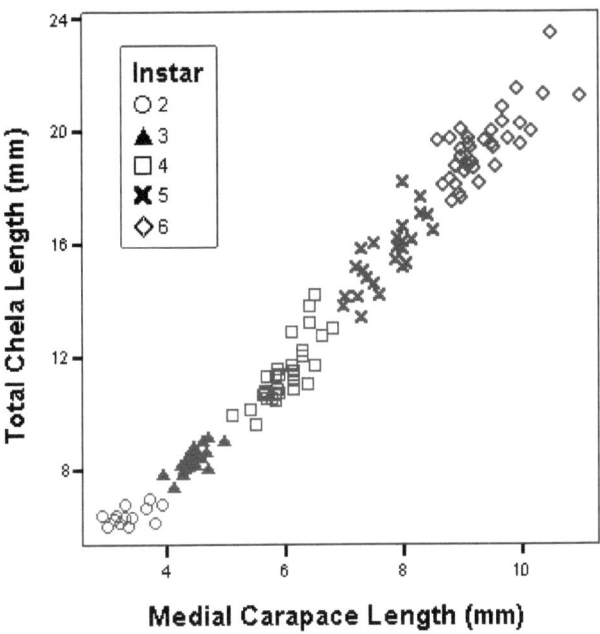

Fig. 3. The relationship between medial carapace length (MCaL) and total chela length (ClBL+ClTL) showing distinct groups (instars) of male *Urodacus elongatus* scorpions.

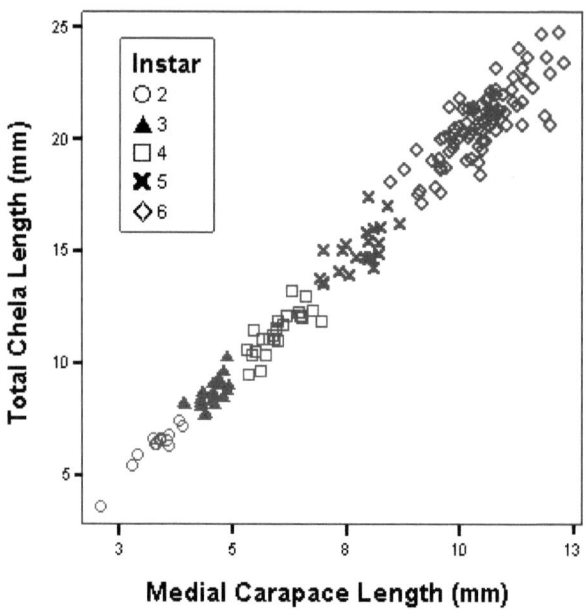

Fig. 4. The relationship between medial carapace length (MCaL) and total chela length (ClBL+ClTL) showing

distinct groups (instars) of female *Urodacus elongatus* scorpions.

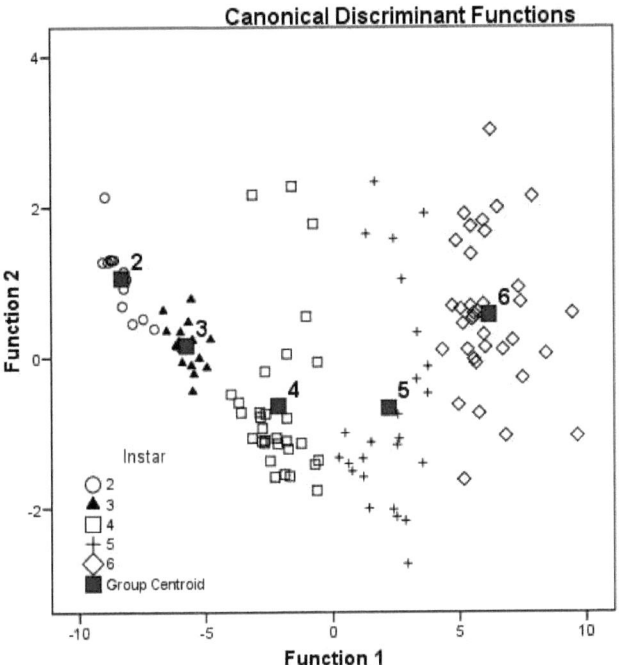

Fig. 5. Male instars identified from a priori groups of *Urodacus elongatus* scorpions in relation to the two significant canonical discriminant functions.

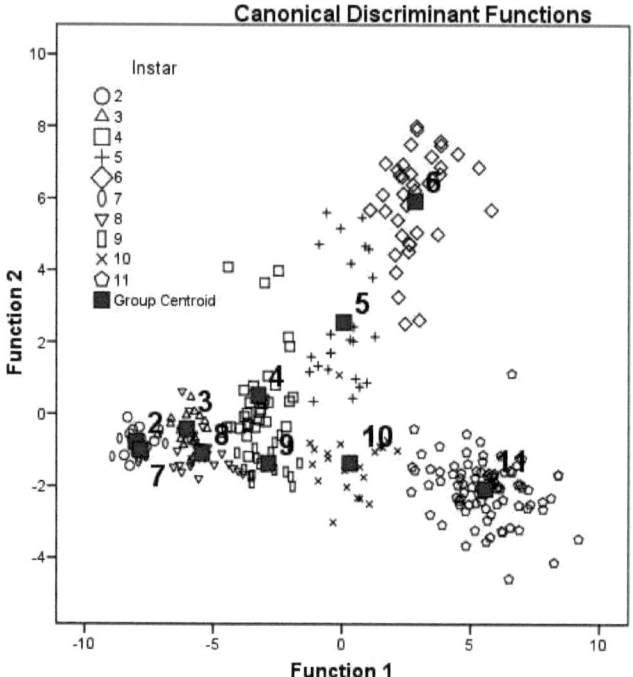

Fig. 6. Instars of both sexes of *Urodacus elongatus* scorpions identified from a priori groups, in relation to the first two significant canonical discriminant functions.

Instars 2-6=male and 7-11=female.

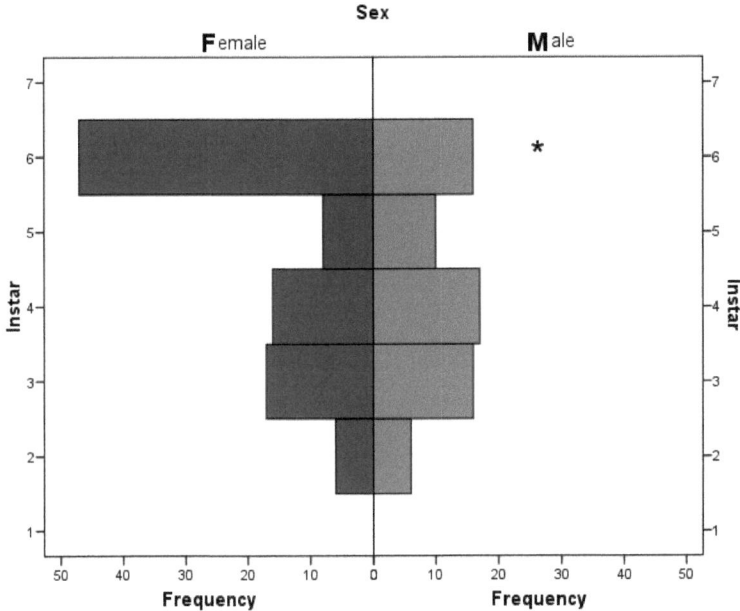

Fig. 7. Population tower of capture frequencies for instars of each sex of *Urodacus elongatus* scorpions. * = significant intersexual difference.

APPENDIX 1

Scorpion Recapture Analysis

Data were entered into Microsoft Excel to create a calendar of catches (Appendix 2). Jolly-Seber Method analysis of recaptures was performed (see next page). This analysis produced population size estimates (Ni) and standard errors (S.E.i) that were extremely variable. Thus, no useful additional information was provided by the analysis and it was not included in the final results. Future work on *U. elongatus* should incorporate a larger search area and sample size in order for the analysis to produce meaningful results. Searches should encompass a standardised, repeatable method of scanning the study area to ensure the entire population is sampled and not limited to active individuals.

Table 5. The recapture data and population estimates derived using the Jolly-Seber Method for a population of *Urodacus elongatus* sampled March to August 2008. n_i = sample size; R_i = scorpions marked and released; r_i = scorpions of the R_i release subsequently recaptured; m_i = marked scorpions in sample; Z_i = scorpions marked before ith sample, not recaptured in ith sample but were subsequently recaptured; a_i = proportion of marked scorpions in population at ith sampling; M_i = marked scorpions in population immediately preceding ith sample; N_i = population size; S.E.$_i$ = standard error of population size; p_i = probability that scorpion alive at release of ith sample will survive and not emigrate before i+1 sample; A_i = scorpions joining population by birth or immigration between ith and i+1 samples, and still alive at i+1 sample.

Sample Dates	n_i	R_i	r_i	m_i	Z_i	a_i	M_i	N_i	S.E.$_i$	p_i	A_i
7-9/ 03/2008	22	22	16	1	n/a	n/a	n/a	n/a	n/a	n/a	n/a
15-16/ 03/2008	10	8	3	1	15	0.1	41	410	433.13	0.55	-66.25
22-23/ 03/2008	6	6	4	3	17	0.17	26.5	159	150.61	0.57	70.30
5-6/ 04/2008	27	23	28	11	18	0.11	17.79	160.07	77.20	1.17	-21.90
4-5/ 05/2008	40	40	42	17	35	0.28	44.33	161.21	35.35	2.12	-3.41
20-21/ 05/2008	37	37	16	9	60	0.46	155.75	338.99	85.75	1.23	328.44
6-7/ 06/2008	31	31	10	18	67	0.29	216.7	746.41	286.27	0.90	-301.56
1-2/ 07/2008	31	30	9	16	59	0.58	214.67	369.70	114.90	2.71	-116.13
18-19/ 07/2008	23	23	2	20	52	0.70	614	882.63	603.95	0.51	-38.17
1-2/ 08/2008	26	26	3	37	34	0.77	314.67	409.07	221.12	n/a	n/a
30-31/ 08/2008	48	47	n/a	n/a	n/a	n/a	n/a	n/a	n/a	n/a	n/a

#	Instar	Sex	7/03/2008	8/03/2008	9/03/2008	15/03/2008	16/03/2008	23/03/2008	5/04/2008	6/04/2008	4/05/2008	5/05/2008	20/05/2008	21/05/2008	6/06/2008	7/06/2008	1/07/2008	2/07/2008	18/07/2008	19/07/2008	1/08/2008	2/08/2008	30/08/2008	31/08/2008
20	2	M	x																					
21	5	M	x																					
22	5	M	x			x			x															
23	6	M	x																					
24	6	F	x																					
25	6	F	x								x												x	
26	6	F	x																					
27	6	M	x							x								x				x		x
28	6	F	x										x										x	
29	4	M		x																				
30		U		x																				
31		U		x																				
32		U		x																				
33	6	M		x																				
34	6	M		x	x																			
35	6	M		x																				
36	6	F		x									x				x							x
37	6	F			x			x				x					x							
38	4	M		x																				
39	6	M		x																				
40	3	F		x																				
41	4	M			x																			
43	6	F			x					x	x	x												
44	6	F			x																			
45	6	F			x																			
46	6	M				x																		
47	4	M				x																		
48	5	M				x																		
49		U					x																	
50	3	M					x																	x
51	4	M					x															x		
52	6	F					x	x																
53	4	F					x									x								
54	3	F						x																
56	6	F						x	x	x		x							x					
57	4	F						x						x			x					x		x
59	6	F						x		x												x		
60	3	F						x				x										x	x	

29

ID	n	Sex	1	2	3	4	5	6	7	8	9	10	11	12	13	14	15
62	3	F	x														
63	6	M	x														
64	6	F	x			x		x			x				x		
65	6	F	x		x		x										
66	6	M	x				x										
67	2	F		x													
68	2	F		x													
69	2	F		x													
70	3	M		x													
72	4	F		x													x
73	3	M		x						x							
74	6	F	x				x			x			x		x		
75	6	F	x														
76	6	F	x	x	x	x											
79	6	F			x												
80	2	F			x												
81	6	M			x					x					x	x	
82	5	F			x												
83	6	F			x												
84	6	F			x		x					x		x	x	x	
85	6	F			x												
86	6	F			x												
87	3	M			x												
88	6	F			x												
89	6	F			x				x					x	x	x	
90	5	M			x		x		x					x			
91	5	F			x												
92	5	F			x											x	
93	6	F			x		x							x	x		x
94	6	F			x			x		x				x	x		
95	6	F			x		x										
98	4	F				x	x					x			x		
99	4	M				x											
100	3	F				x											
101	2	M				x											
102	2	M				x											
103	2	M				x											
104	5	F				x	x										
105	5	M				x			x			x			x		
106	6	F				x							x				
107	6	F				x	x										
108	6	M				x	x							x	x		
109	6	F				x	x			x				x	x	x	
110	4	M				x										x	
111	4	F				x											

ID	N	Sex	1	2	3	4	5	6	7	8	9	10	11
112	4	F	x										
113	6	F	x						x				x
114	5	M	x										
115	6	M	x										
116	6	F	x				x					x	
117	3	F		x									
118	4	M		x									
119	4	M		x			x						
120	3	F		x									
121	3	F		x									
122	3	M		x									
123	5	F		x									
124	5	M		x	x								
125	6	F		x			x			x	x		
126	6	M		x	x						x		
127	6	F		x	x								
128	6	F		x									
129	6	F		x				x		x		x	
130	2	M			x								
131	2	F			x								x
132	3	M			x								
133	4	F			x								
134	4	M			x								
135	3	M			x								
136	4	F			x								
137	2	M			x								
138	5	F			x		x						
139	4	M			x						x	x	
140	5	M			x								
141	3	M			x								
142	3	F			x				x				
143	5	F			x								
144	4	M			x		x						
145	6	F			x			x					x
146	6	F			x								
147	3	M				x							
148	4	F				x			x				
149	4	F				x							
150	6	F				x						x	
151	6	F				x							
152	2	M					x						
153	3	F					x						
154	2	M					x						
155	4	M					x				x		x
156	3	F					x						x

ID			C1	C2	C3	C4	C5	C6	C7	C8	C9	C10	C11	C12	C13	C14	C15	C16	C17
157	6	F											x		x				
158	2	M												x					
159	2	M												x					
160	2	F												x			x		
161	2	F												x					x
162	2	F												x					x
163	3	M												x	x		x		
164	3	F												x					
165	5	F													x		x		
166	6	F													x		x		
167	2	F													x				
168	3	M													x				
169	4	M													x				
170	4	F														x			
171	4	M														x			
172	6	F														x			
173	5	M														x		x	
174	6	F														x	x		
175	4	F															x	x	
176	3	M														x			
177	3	F														x			
178	4	F														x			
179	3	M														x			
180	4	M														x			
181	6	M														x			
183	3	F														x			
184	3	F																	x
185	3	M																	x
186	6	F																	x
187	6	F																	x
42D	4	M			x														
45.5D		U			x														
55D	8	M					x												
58P	6	F					x										x		
61D	3	F					x												
71D	2	F						x											

Habitat Requirements and Population Density of the Scorpion *Urodacus elongatus* (Scorpionidae)

INTRODUCTION

The Flinders Ranges in South Australia spans a region of biogeographical transition from the more arid north to the more temperate south (Davies *et al.* 1996). Within the Ranges the location of Mount Remarkable National Park is particularly biogeographically significant as the flora and fauna present include elements typical of both climates (Davies *et al.* 1996). Current biota is likely to be either opportunistic species that have adapted and evolved within the patches of suitable habitat, or relict species with fragmented habitat. Reproductive isolation of such populations has lead to the evolution of species, endemic to the region. Fragmentation of habitat regions like the Flinders Ranges can increase susceptibility of resident species to disturbance. The degree of disturbance impact closely relates to habitat specificity of the species. Therefore, conservation and management of species in the Flinders Ranges requires a thorough understanding of their specific habitat requirements.

A major habitat requirement is shelter, and the nature of shelter use relates to the behaviour and physiology of the species. Some species are more active foragers and continuously move through their habitat, utilising temporary shelters. Others are more sedentary sit-and-wait foragers and spend most of their time in specific areas of their habitat, and shelters need to be more permanent and secure. The influences of the various components of the habitat will be very different for these two groups. More transient shelter types like vegetation and litter will be more important to active foragers. In contrast, more permanent shelter structures like rocks and burrows will be more important to sedentary sit-and-wait foragers. The presence of habitat components can have a positive association (Bradley 1986; Fet 1980; Hofer *et al.* 1996; Short *et al.* 2001; Trainor *et al.* 2000), or negative association (Bradley 1986; Short *et al.* 2001; Woods and Schiel 1997) with population density of resident species. The nature of these associations will depend on how the species habitat selection is influenced by the structure of the habitat.

Habitat selection is also influenced by climate and circadian patterns of activity. In hot, dry habitats species will use behaviours to tolerate or avoid over-heating and dehydration (Downes and Shine 1998a; b; Huey *et al.* 1989; Kearney 2002). Avoidance behaviours often include burrowing or being nocturnally active (Kearney 2002; Pye *et al.* 1999; Shah *et al.* 2004; Short *et al.* 2001; Woods and Schiel 1997). Scorpions live in hot, dry habitats and, being ectotherms, they have evolved such avoidance behaviours to assist with homeostasis (Anderson 1983; Polis and Farley 1980; Warburg 2000). A flattened, elongated, segmented body shape renders scorpions highly successful at finding shelter in crevices, under rocks, or burrowing into the sediment. Although largely covered by a relatively impermeable exoskeleton, inter-segmental body surfaces have only a membranous cover which is quite permeable to water (Gefen and Ar 2004; Polis 1990). Understanding the biology and ecology of scorpions enhances our ability to conserve and manage their continued presence within their natural habitats.

Scorpions are an ancient group of arachnids historically represented in folklore and mythology. Human perception of the combination of four pairs of hairy legs, large pedipalps, and "tail" (metasoma) with a venomous telson has caused them to be feared and reviled. Previous research on scorpions has been limited to studies of natural histories, sensory physiology, and venom toxicology papers (Brown 2004; Carthy 1968; Lourenco *et al.* 2003; Ozkan and Carhan 2008; Yigit and Benli 2008). The biology of scorpions has been well established, and the ecology of scorpions has been studied in numerous northern hemisphere species (Benton 1992; Bradley 1984; Bradley 1986; 1988; Fet 1980). However, the southern hemisphere species, and in particular the subfamily Urodacinae remain poorly understood.

The subfamily Urodacinae is endemic to Australia, and distribution of *Urodacus* species ranges from widespread (e.g. *U. yaschenkoi,* and *U. manicatus*) to highly isolated (e.g. *U. centralis,* and *U. spinatus*) (Koch 1977). Research into *Urodacus* scorpions is scant and most occurred decades ago (Shorthouse and Marples 1982; Smith 1966). This present study is the first to focus on a relictual species of *Urodacus* with a limited geographic range, and provides new information that will assist conservation of this, and other similarly relictual species.

The endemic Flinders Ranges scorpion (*Urodacus elongatus*) (Scorpionidae) shows a patchy distribution. *U. elongatus* scorpions are sit-and-wait predators that construct short burrows and scrapes underneath rocks (J. Munro, pers. obs., 2008). Other scorpions also dig burrows under rocks or similar permanent structures (Benton 1992; Bradley and Brody 1984). Some dig spiral burrows in open ground (Bradley 1986; Fet 1980; Polis 1980). Poor eyesight causes scorpions to be reliant on mechano-receptive detection of vibration and air movement via sensilla (slits) and setae (hairs) as their primary sensory input (Anderson 1983; Polis 1990). Reliance by scorpions on such methods of sensory detection enables easy capture. This renders them susceptible to harvesting for the pet trade, the impacts of which remain unknown, and highlights the need for this study. The aims of this study are to 1) identify the habitat variables that predict the distribution of *U. elongatus*, and 2) quantify the population density of *U. elongatus* in suitable habitat.

METHODS

Study Sites

This study was done in the southern Flinders Ranges at the Mount Remarkable National Park (MRNP) (32°46'38.20 S 138°04'30.00 E) and Telowie Gorge Conservation Park (TGCP) (32°58'58.00 S 138°05'35.40 E) (Fig. 1). The MRNP ranges from the western coastal plain near Mambray Creek to the summit of Mount Remarkable in the east, incorporating the Alligator and Mambray Creek catchments. The TGCP is a smaller area approximately 30km south of MRNP and incorporates the Telowie Creek catchment. Vegetation present at both sites is primarily mallee woodland with some bare scree slopes (D.E.H. 2001; Davies *et al.* 1996). The predominant flora includes *Eucalyptus, Callitris glaucophylla* and *Acacia*, native and introduced grasses, and low-growing herbaceous plants (D.E.H. 2001; Davies *et al.* 1996). Rocks are quartzite and sandstone. Soils are generally alkaline, well-drained, reddish, dense loams, shallow on slopes and deeper in gullies (D.E.H. 2001; Davies *et al.* 1996). The regional climate is Mediterranean, with long, hot dry summers (average maximum temperature 30°C), and cool, wet winters (average maximum temperature 20°C). The majority of rainfall occurs from April to October, with variable spatial distribution due to local topography. Regional annual rainfall averages range from 347mm at Mambray Creek in the west to 589mm at Melrose, in the east of MRNP (D.E.H. 2001).

Habitat Survey

Habitat data were collected between February and July 2008. Due to the range and variability of habitat site qualities throughout MRNP and the restriction on access due to topography, a stratified sampling technique was employed. Sites were located at randomly selected distances from fire-tracks and paths within MRNP (Fig. 1 and Appendix 1) and incorporated ridges, slopes, and gullies. Each site was a 5x5m quadrat that was photographed and located with GPS. Tree cover was visually estimated as percent coverage and identified by genus. Groundcover was also estimated as percent coverage and categorised as: rock; vegetation; litter; and bare ground. Slope was visually estimated and aspect was obtained from compass bearings. Definitions of measurements recorded for each site are outlined in Table 1.

Rock Qualities

The habitat survey identified rock cover as an important predictor of scorpion presence (see results). To determine whether particular rock qualities were preferred by scorpions, I did a follow-up study. All rocks within each 5x5m habitat site were lifted and scorpion presence beneath was recorded. If scorpions were present rock length, width, and depth (mm) were measured, and rock volume (m^3) was calculated. Rock volume was defined as length x width x depth (see Table 1 for definitions), and recorded on the scale: 0.010-$0.050m^3$(1), 0.051-$0.100m^3$(2), 0.101-$0.150m^3$(3), 0.151-$0.200m^3$(4), 0.201-$0.250m^3$(5), 0.251-$0.300m^3$(6), $>0.300m^3$(7). This procedure was repeated for three randomly selected rocks within each site, which may or may not have harboured scorpions. Rocks were randomly selected using a random number table (Zar 1984) to identify predetermined x and y co-ordinates within quadrats.

Population Density

Scorpions found were placed individually into zip-lock bags using 30cm forceps. The scorpion instar was identified using the criteria developed in a concurrent study (J. Munro, unpublished manuscript, 2008).

Statistical Analysis

Raw data were entered into a Microsoft Office Excel 2007 software program. Values of percent coverage that were less than five were converted to a value of 2.5 for inclusion in statistical analysis.

Habitat Characteristics

A multiple linear regression was performed to identify which of the measured variables (Table 1) predicted scorpion abundance. ANOVA was used to test the significance of the regression.

A multiple logistic regression was performed to analyse the presence/absence data for *U. elongatus* in relation to the same environmental variables. Forward stepwise method was selected as this allowed for identification of individual variables that showed a significant influence, to be included in a reduced model.

Rock Characteristics

A second multiple logistic regression was performed to examine the influences of rock volume, length, width, and depth on the presence/absence of scorpions. Forward stepwise method was again selected to identify the rock qualities contributing significantly to the model ($P<0.001$).

Hosmer and Lemeshow statistics were used to test if the reduced logistic regression model improved the goodness-of-fit of the model to the data, compared with the original model, in both logistic regressions (Hair *et al.* 1995; Zar 1984).

Population Density

Number of scorpions per 15x15m site was used to calculate scorpion density per hectare. The frequency distribution of population densities was tested for random expectations derived from the Poisson distribution, using the chi-square statistic.

All statistical analysis was performed using SPSS 15.0 for Windows.

RESULTS

Habitat Characteristics

Only 28% (17 of 61 sites) of quadrats were inhabited by *Urodacus elongatus*. Total number of scorpions present per site ranged from one to five. Multiple linear regression was not significant (one-way ANOVA: $F_{8,52}$=1.532, *P*=0.17), and explained only 19% (R^2) of the variance in the data. However the multiple logistic regression analysis indicated that the presence or absence of scorpions can be predicted from the percentage of rock cover and the degree of slope (Table 2). Odds ratios from the logistic regression indicated that percentage rock cover and slope were the most important variables influencing the presence or absence of scorpions (Table 2). Hosmer and Lemeshow statistics showed that the reduced model improved the goodness-of-fit of the model to the data, compared with the original model (Hair *et al.* 1995; Zar 1984). The likelihood of finding scorpions tended to increase with percentage of rock cover, with the greatest likelihood in the 21-40% range (Fig. 2). In contrast, the likelihood increased with slope to a plateau in the 11-30° range, after which scorpions were absent (Fig. 3).

Rock Characteristics

The presence of scorpions with relation to rock can be accurately predicted from a linear combination of rock volume, rock width, and rock depth (Table 3). The likelihood of finding scorpions under the rocks tended to produce an approximately normal pattern of distribution, for all three predictor variables (Fig. 4). Odds ratios from the logistic regression indicated that rock width and rock depth were the most important rock size variables influencing the presence of scorpions (Table 3). Hosmer and Lemeshow statistics showed that the reduced model improved the goodness-of-fit of the model to the data, compared with the original model (Hair *et al.* 1995; Zar 1984).

Population Density

Overall site population density of scorpions varied from 0 to 667/ha (mean = 84.5±19.2/ha (mean ±S.E.), *n*=61). Sites without scorpions were twice as frequent as any category of sites with scorpions present (Fig. 5). When the sites where scorpions were absent were excluded, this overall site population density increased to 161.1±31.1/ha (mean ±S.E.) over the range

44-667/ha, n=32. Frequency distribution of scorpion densities was not significantly different from random expectations, as predicted by the Poisson distribution (X^2= 0.00, P>0.9).

DISCUSSION

This study examined the association between environmental variables and the presence/absence/abundance of scorpions. Scorpion abundance could not be significantly predicted by any of the environmental variables measured. However presence/absence of scorpions was able to be predicted from, rock cover and slope (Table 2). In terms of specific rock characteristics, scorpions were most often found under medium-sized rocks (Fig. 4).

U. elongatus were associated with slightly sloping sites (11-30°) (Fig. 3). In high rainfall areas, scorpion density has been positively related to slope, where steep, well-drained areas resist flooding (Bradley 1986). However, infrequent rainfall and absence of scorpions on steeper slopes in this study indicate the significance of slope is more likely due to a covariance with rock presence.

U. elongatus shelter under rocks from predators such as reptiles and birds. They also use their burrow entrance as sit-and-wait ambush style predators. The presence of *U. elongatus* was significantly positively associated with rock, and particularly moderate (21-40%) rock coverage (Fig. 2). This indicates, sites with moderate rock cover with the right rock characteristics (e.g. creek rocks), and lower slopes (e.g. floodplains or creeks), provide suitable habitat for scorpions. Lower scorpion presence in sites with low rock cover is likely due to avoidance of, over-exposure to predators when active, and increased soil temperatures of bare ground. Lower presence in high rock cover, may reflect reduced manoeuvrability, or insufficient soil for burrowing.

Distribution and abundance can also increase with habitat structural complexity. Increased potential resource niches, e.g. shelter, are able to support higher population density, potentially improving reproductive success (Roediger and Bolton 2008). After the first moult the young of *U. elongatus* become independent (J. Munro, unpublished data, 2008). Searching time and distance, to find a suitable rock, will vary according to rock cover and characteristics adjacent to the mother's rock. Small, naive scorpions are susceptible to high mortality due to encounters with predators, during rock selection and settlement. Predators include larger scorpions, and conspecific cannibalism tends to decline with increased variation in diurnal patterns of activity between age groups of scorpions (Polis 1980). Competition for shelter between conspecific adult and juvenile scorpions can strongly influence the quality of rocks used by both groups. Previous study on rock-sheltering crustaceans has demonstrated that larger, more competitive individuals maintain residence in the best quality shelters (Dennis *et al.* 1997; Martin and Moore 2007).

U. elongatus presence was significantly associated with rock size, especially at median rock volume ($0.151-0.200m3$), width and depth (Fig. 4). This suggests medium sized rocks represent more suitable habitat than smaller or larger rocks. Lower scorpion presence under smaller rocks could relate to exposure (i.e. thermal conductance, insufficient shading, or predation). Firstly, exposure threats in scorpions are worsened by desiccation due to small surface-to-volume ratios, and transpiration via exposed inter-segmental membranes. Although scorpions can tolerate dry conditions (Gefen and Ar 2004; Margules and Milkovits 1994), many species do this by using avoidance behaviour, such as burrowing to prevent desiccation (Benton 1992; Bradley 1986; Bradley and Brody 1984; Fet 1980; Polis 1980). Secondly, thermal benefits of shade, particularly during midday temperatures, can distinctly impact species distribution and behaviour in their habitat (Cain *et al.* 2008; Kearney 2002; Shah *et al.* 2004). Burrowing and sheltering under rocks, and nocturnal activity, are avoidance behaviours used by scorpions against solar radiation (Anderson 1983; Polis and Farley 1980; Warburg 2000). Finally, poor eyesight and a tendency to fight rather than flight (J. Munro, pers. obs., 2008), renders *U. elongatus* scorpions under small rocks, more vulnerable to diurnal attack by predators using visual prey detection.

Simultaneously, lower use of larger rocks by scorpions could be due to insufficient thermal conductance. Absorption of solar radiant energy by rock during daylight hours provides warmth via conductance, particularly beneficial to ectotherms (Downes and Shine 1998a; b; Huey *et al.* 1989; Shah *et al.* 2004). Lower conductance in large rocks may reduce the diurnal and seasonal tolerance of scorpions to above-ground foraging, during colder periods. Additionally, larger rocks may release less heat after sunset during warmer months when nocturnal invertebrate prey are most available, reducing daily foraging efficiency. Lower scorpion presence under larger rocks is not likely to be an artefact of maximum rock size rolled since presence was normally distributed over rock sizes sampled, indicating a decline was already established within the sampled range (Fig. 4).

Scorpion presence was not significantly associated with aspect, tree cover, vegetation, litter, or bare ground (Table 2). The results for aspect may reflect the nocturnal activity and burrowing of scorpions, both rendering them less affected by the patterns of sunlight and weather. The absence of significant associations of scorpion presence with the vegetation or litter scales measured highlights the need for future assessment at a finer scale, to more accurately define any associations present. Future work should specifically target litter, as this is the scale of vegetation with which the scorpions are most likely to be interacting.

Conclusions

This study has shown that rock and slope were the significant predictors of the presence of *U. elongatus* scorpions. In MRNP and TGCP scorpions occur on slightly sloping, moderately rocky creek beds and floodplains. Within the Parks these habitats represent areas most impacted by humans, either by tourist activity or residual impacts from adjacent farming practices. Thus, these habitats are most at risk of disturbance. Conservation and management of *U. elongatus* will require preservation and protection of these essential habitat areas on a local scale in the Parks, and more widely throughout the Flinders Ranges.

REFERENCES

Anderson R. C. (1983) *Scorpions the ancient arachnids*. Idaho Museum of Natural History.

Benton T. G. (1992) The ecology of the scorpion *Euscorpius flavicaudis* in England. *Journal of Zoology, London* **226**, 351-68.

Bradley R. A. (1984) The influence of the quantity of food on fecundity in the desert grassland scorpion (*Paruroctonus utahensis*) (Scorpionida, Vaejovidae): An experimental test. *Oecologia* **62**, 53-6.

Bradley R. A. (1986) The relationship between population density of *Paruroctonus utahensis* (Scorpionida: Vaejovidae) and characteristics of its habitat. *Journal of Arid Environments* **10**, 000-.

Bradley R. A. (1988) The influence of weather and biotic factors on the behaviour of the scorpion (*Paruroctonus utahensis*). *Journal of Animal Ecology* **57**, 533-51.

Bradley R. A. & Brody A. J. (1984) Relative abundance of three Vaejovid scorpions across a habitat gradient. *Journal of Arachnology* **11**, 437-40.

Brown C. A. (2004) Life histories of four species of scorpion in three families (Buthidae, Diplocentridae, Vaejovidae) from Arizona and New Mexico. *Journal of Arachnology* **32**, 193-207.

Cain J. W., Jansen B. D., Wilson R. R. & Krausman P. R. (2008) Potential thermo regulatory advantages of shade use by desert bighorn sheep. *Journal of Arid Environments* **72**, 1518-25.

Carthy J. D. (1968) The pectines of scorpions. *Symposium of the zoological society of London* **23**, 251-61.

D.E.H. (2001) *Mount Remarkable National Park Management Plan - Draft*. Department for Environment and Heritage, Adelaide, South Australia.

Davies M., Twidale C. R. & Tyler M. J. (1996) *Natural History of the Flinders Ranges*. Royal Society of South Australia Inc., Adelaide.

Dennis D. M., Skewes T. D. & Pitcher C. R. (1997) Habitat use and growth of juvenile ornate rock lobsters, *Panulirus ornatus* (Fabricius, 1798), in Torres Strait, Australia. *Marine and Freshwater Research* **48**, 663-70.

Downes S. & Shine R. (1998a) Heat, safety or solitude? Using habitat selection experiments to identify a lizard's priorities. *Animal Behaviour* **55**, 1387-96.

Downes S. & Shine R. (1998b) Sedentary snakes and gullible geckos: predator-prey co evolution in nocturnal rock-dwelling reptiles. *Animal Behaviour* **55**, 1373-85.

Fet V. Y. (1980) Ecology of the scorpions (Arachnida, Scorpiones) of the southeastern Kara-Kum. *Entomologicheskoe Obozrenie* **59**, 165-70.

Gefen E. & Ar A. (2004) Comparative water relations of four species of scorpions in Israel: evidence for phylogenetic differences. *The Journal of Experimental Biology* **207**, 1017-25.

Hair J. F., Anderson R. E., Tatham R. L. & Black W. C. (1995) *Multivariate Data Analysis*. Prentice-Hall Inc., New Jersey.

Hofer H., Wollscheid E. & Gasnier T. (1996) The relative abundance of *Brotheas Amazonicus* (Chactidae, Scorpiones) in different habitat types of a central Amazon rainforest. *The Journal of Arachnology* **24**, 34-8.

Huey R. B., Peterson C. R., Arnold S. J. & Porter W. P. (1989) Hot Rocks and Not-So-Hot Rocks: Retreat-Site Selection by Garter Snakes and Its Thermal Consequences. *Ecology* **70**, 931-44.

Kearney M. (2002) Hot rocks and much-too-hot rocks: seasonal patterns of retreat-site selection by a nocturnal ectotherm. *Journal of Thermal Biology* **27**, 205-18.

Koch L. (1977) The taxonomy, geographic distribution and evolutionary radiation of Australo-Papuan scorpions. *Records of the Western Australian Museum* **5**, 83-367.

Lourenco W. R., Andrzejewski V. & Cloudsley-Thompson J. L. (2003) The life history of *Chactas reticulatus* (Kraepelin, 1912) (Scorpiones, Chactidae), with a comparative analysis of the reproductive traits of three scorpion lineages in relation to habitat. *Zoologischer Anzeiger* **242**, 63-74.

Margules C. R. & Milkovits G. A. (1994) Contrasting effects of habitat fragmentation on the scorpion *Cercophonius squama* and an amphipod. *Ecology* **75**, 2033-42.

Martin A. L. & Moore P. A. (2007) Field observations of agonism in the crayfish, *Orconectes rusticus*: shelter use in a natural environment. *Ethology* **113**, 1192-201.

Ozkan O. & Carhan A. (2008) The neutralizing capacity of *Androctonus crassicauda* antivenom against *Mesobuthus eupeus* scorpion venom. *Toxicon* **52**, 375-9.

Polis G. A. (1980) Seasonal Patterns and Age-Specific Variation in the Surface Activity of a Population of Desert Scorpions in Relation to Environmental Factors. *The Journal of Animal Ecology* **49**, 1-18.

Polis G. A. (1990) *The Biology of Scorpions*. Stanford University Press, Stanford.

Polis G. A. & Farley R. D. (1980) Population biology of a desert scorpion: Survivorship, microhabitat, and the evolution of life history strategy. *Ecology* **61**, 620-9.

Pye T., Swain R. & Seppelt R. D. (1999) Distribution and habitat use of the feral black rat (*Rattus rattus*) on subantarctic Macquarie Island. *Journal of Zoology* **247**, 429-38.

Roediger L. M. & Bolton T. F. (2008) Abundance and distribution of South Australia's endemic sea star, *Parvulastra parvivipara* (Asteroidea: Asterinidae). *Marine and Freshwater Research* **59**, 205-13.

Shah B., Shine R., Hudson S. & Kearney M. (2004) Experimental analysis of retreat-site selection by thick-tailed geckos *Nephrurus milii*. *Austral Ecology* **29**, 547-52.

Short F. T., Matso K., Hoven H. M., Whitten J., Burdick D. M. & Short C. A. (2001) Lobster use of eelgrass habitat in the Piscataqua River on the New Hampshire/Maine border, USA. *Estuaries* **24**, 277-84.

Shorthouse D. J. & Marples T. G. (1982) The life stages and population dynamics of an arid zone scorpion (*Urodacus yaschenkoi*) (Birula 1903). *Austral Ecology* **7**, 109-18.

Smith G. T. (1966) Observations on the life history of the scorpion *Urodacus abruptus* (Scorpionidae), and the analysis of its home sites. *Australian Journal of Zoology* **14**, 383-98.

Trainor C., Fisher A., Woinarski J. & Churchill S. (2000) Multiscale patterns of habitat use by the Carpentarian rock-rat (*Zyzomys palatalis*) and the common rock-rat (*Z. argurus*). *Wildlife Research* **27**, 319-32.

Warburg M. R. (2000) Intra- and interspecific cohabitation of scorpions in the field and the effect of density, food, and shelter on their interactions. *Journal of Ethology* **18**, 59-63.

Woods C. M. C. & Schiel D. R. (1997) Use of seagrass *Zostera novazelandica* (Setchell, 1933) as habitat and food by the crab *Macrophthalmus hirtipes* (Heller, 1862) (Brachyura: Ocypodidae) on rocky intertidal platforms in southern New Zealand. *Journal of Experimental Marine Biology and Ecology* **214**, 49-65.

Yigit N. & Benli M. (2008) The venom gland of the scorpion species *Euscorpius mingrelicus* (Scorpiones: Euscorpiidae): morphological and ultrastructural characterization. *Journal of Venomous Animals and Toxins Including Tropical Diseases* **14**, 466-80.

Zar J. (1984) *Biostatistical Analysis, 2nd Ed.* Prentice-Hall Inc., New Jersey.

Table 1. Ecological variables recorded for each habitat site. Definitions and measurement scales used are outlined.

Predictor Variable	Measurement and Definition
Slope	Estimated to nearest 5°
Aspect	Estimated to nearest 45°
Tree Cover	Percent coverage 0-100% (estimated to nearest 5%) of projected foliage >1m
Ground Cover	Percent coverage (estimated to nearest 5%) to a combined to total 100% incorporating:
	Rock, diameter >10cm
	Vegetation = Projected foliage <1m
	Litter = Dead plant matter
	Bare Ground = Substratum free from cover
Rock Length	Longest horizontal axis (mm)
Rock Width	Longest horizontal axis perpendicular to length (mm)
Rock Depth	Longest vertical axis (mm)

Table 2. Multiple logistic regression of the presence or absence of *Urodacus elongatus* in relation to two predictor variables, percentage rock cover and slope [Logit P = -1.972– (0.082 x rock cover)-(0.080 x slope)]. B = estimated model coefficients, S.E. = standard error of B.

Note that other non-significant predictor variables were removed from the model to improve the fit of the model to the data.

Predictor	B	S.E.	Wald's X^2	P	Odds ratio
Constant	-1.972	0.636	9.630	0.002	0.139
Rock	0.082	0.025	10.839	0.001	1.085
Slope	-0.080	0.032	6.452	0.011	0.923

Test				
Overall model evaluation		X^2	P	
Omnibus Test of Model Coefficients		22.550	<0.001	
-2 x log-likelihood		49.638		
Hosmer and Lemeshow		4.995	0.758	

Table 3. Multiple logistic regression of the presence or absence of *Urodacus elongatus* in relation to three predictor variables, rock volume, rock width and rock depth [Logit P = -3.797– (14.661 x rock volume)+(0.222 x rock width)+(0.186 x rock depth)]. B = estimated model coefficients, S.E. = standard error of B.

Note that other non-significant predictor variables were removed from the model to improve the fit of the model to the data.

Predictor	B	S.E.	Wald's X^2	P	Odds ratio
Constant	- 3.797	0.479	62.934	<0.001	0.022
Rock Volume	-14.661	2.886	25.801	<0.001	0.000
Rock Width	0.222	0.033	43.937	<0.001	1.248
Rock Depth Test	0.186	0.048	15.250	<0.001	1.204

Overall model evaluation	X^2	P
Likelihood ratio test	127.645	<0.001
-2 x log-likelihood	309.900	
Hosmer and Lemeshow	51.397	<0.001

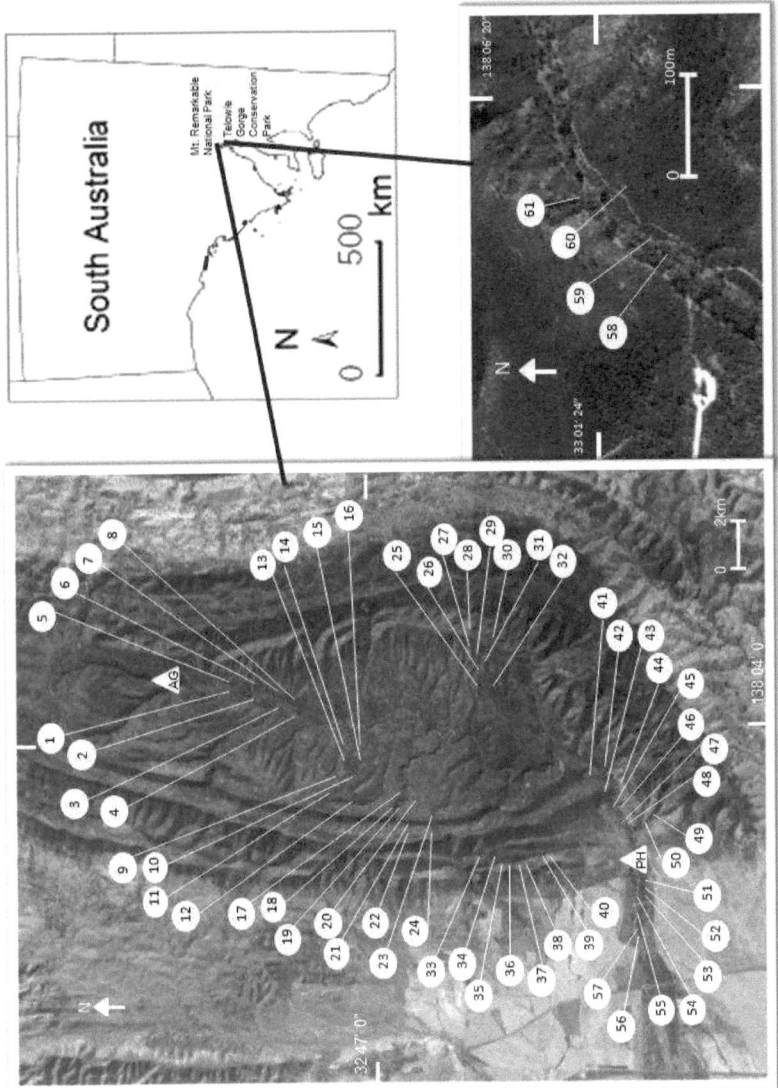

Fig. 1. Location of habitat survey sites in Mount Remarkable National Park and Telowie Gorge Conservation Park, South Australia.

PH= Park Headquarters at Mambray Creek. AG= Alligator Gorge.

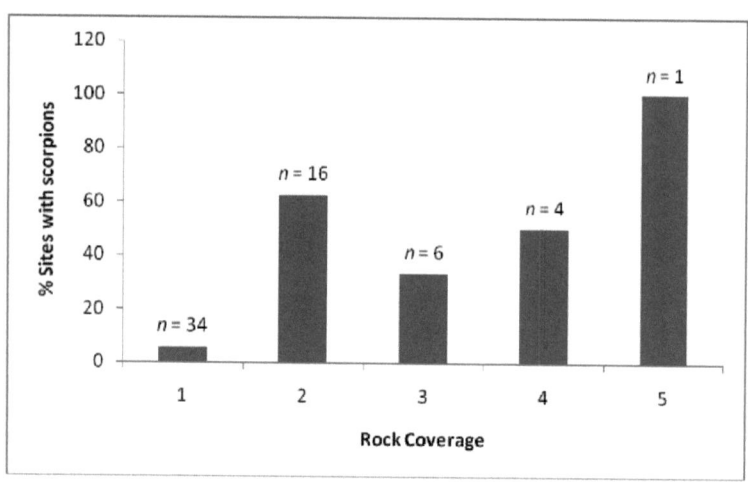

Fig. 2. Habitat sites (5x5m) containing *Urodacus elongatus* scorpions, categorised by percentage of rock coverage: Scale = 0-20%(1), 21-40%(2), 41-60%(3), 61-80%(4), 81-100%(5), *n* = number of sites.

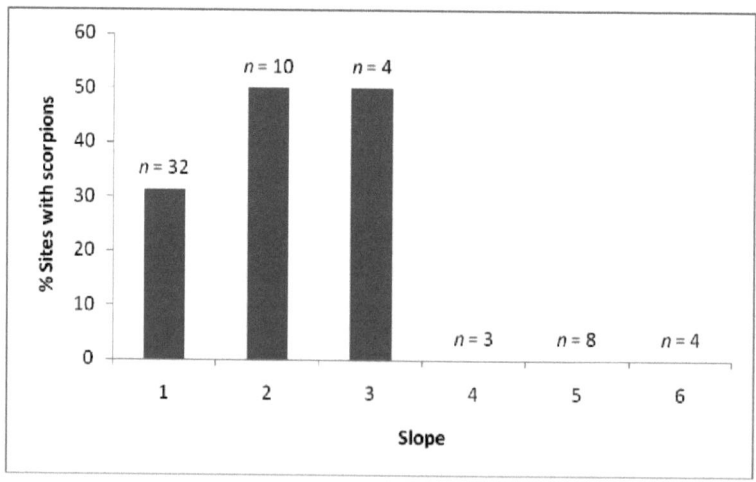

Fig. 3. Habitat sites (5x5m) containing *Urodacus elongatus* scorpions, categorised by slope: Scale = 0-10°(1), 11-20°(2), 21-30°(3), 31-40°(4), 41-50°(5), >50°(6), *n* = number of sites.

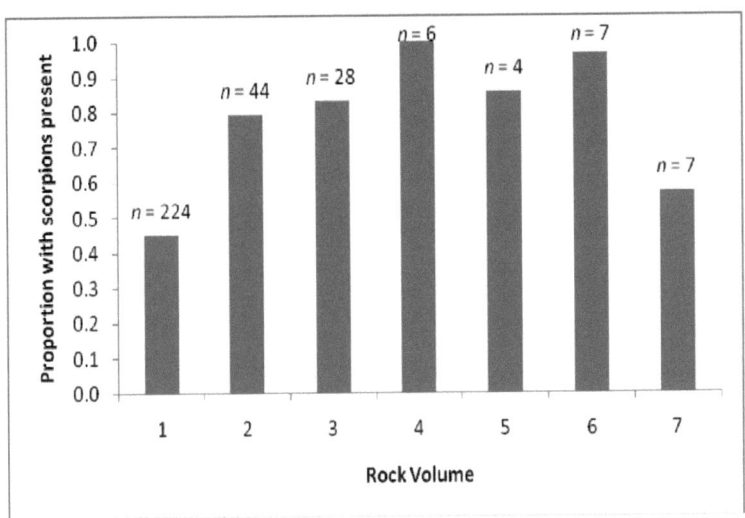

Fig. 4. Proportion of habitat sites (5x5m) containing *Urodacus elongatus* scorpions, categorised by rock volume (m^3): Scale = 0.010-0.050m^3(1), 0.051-0.100m^3(2), 0.101-0.150m^3(3), 0.151-0.200m^3(4), 0.201-0.250m^3(5), 0.251-0.300m^3(6), >0.300m^3(7), *n* = number of rocks.

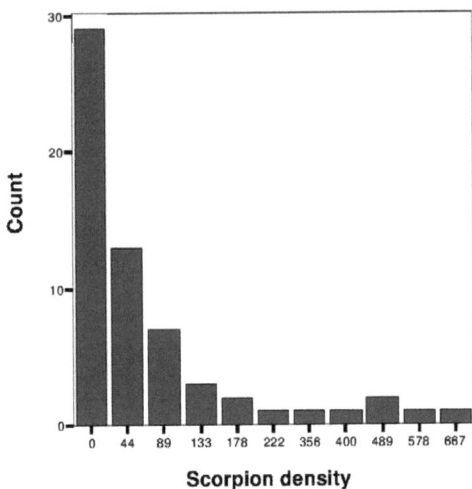

Fig. 5. Counts of the numbers of extended boundary sites (15x15m) as categorised by estimated scorpion density (individuals per ha).

50

APPENDIX 1

Below is a list of the map site reference numbers used in this paper in Fig. 1. Site numbers refer to habitat site names used in the field and in raw data entry. GPS readings relate to the specific position of the site or the respective position on each transect.

Map Site Number	Site	GPS
1	AGH#02	32°45.422' S 138°04.402' E
2	AGH#04	32°45.636' S 138°04.331' E
3	AGH#06	32°45.893' S 138°04.312' E
4	AGH#08	32°46.045' S 138°04.247' E
5	AGH#01	32°45.422' S 138°04.402' E
6	AGH#03	32°45.636' S 138°04.331' E
7	AGH#05	32°45.893' S 138°04.312' E
8	AGH#07	32°46.045' S 138°04.247' E
9	FT#06	32°46.729' S 138°03.187' E
10	FT#05	32°46.729' S 138°03.187' E
11	FT#08	32°46.619' S 138°02.944' E
12	FT#07	32°46.619' S 138°02.944' E
13	FT#04	32°46.768' S 138°03.482' E
14	FT#02	32°46.792' S 138°03.772' E
15	FT#03	32°46.768' S 138°03.482' E
16	FT#01	32°46.792' S 138°03.772' E
17	HG#07	32°47.405' S 138°02.594' E
18	HG#05	32°47.543' S 138°02.511' E
19	HG#06	32°47.543' S 138°02.511' E
20	HG#08	32°47.405' S 138°02.594' E
21	HG#04	32°47.759' S 138°02.512' E
22	HG#03	32°47.759' S 138°02.512' E
23	HG#01	32°47.969' S 138°02.429' E
24	HG#02	32°47.969' S 138°02.429' E
25	MCT#08	32°48.156' S 138°05.382' E
26	MCT#06	N/A
27	MCT#02	32°48.450' S 138°04.715' E
28	MCT#01	32°48.450' S 138°04.715' E
29	MCT#04	32°48.412' S 138°05.083' E
30	MCT#03	32°48.412' S 138°05.083' E

Map Site Number	Site	GPS
31	MCT#05	N/A
32	MCT#07	32°48.156' S 138°05.382' E
33	TB#02	32°48.567' S 138°01.996' E
34	TB#01	32°48.567' S 138°01.996' E

		GPS
35	TB#03	32°48.640' S 138°01.850' E
36	TB#04	32°48.745' S 138°01.861' E
37	TB#05	32°48.883' S 138°01.894' E
38	TB#06	32°49.033' S 138°01.898' E
39	TB#07	32°49.155' S 138°01.932' E
40	TB#08	32°49.279' S 138°01.990' E
41	SGL#07	32°49.867' S 138°03.387' E
42	SGL#08	32°49.867' S 138°03.387' E
43	SGL#05	32°50.039' S 138°03.199' E
44	SGL#06	32°50.039' S 138°03.199' E
45	SGL#03	32°50.080' S 138°02.977' E
46	SGL#04	32°50.080' S 138°02.977' E
47	SGL#01	32°50.146' S 138°02.710' E
48	SGL#02	32°50.146' S 138°02.710' E
49	DG#01	32°50.434' S 138°02.589' E
50	DG#02	32°50.359' S 138°02.541' E
51	MCE#01	32°50.505' S 138°01.454' E
52	MCE#02	32°50.527' S 138°01.644' E
53	MCE#03	32°50.534' S 138°01.627' E
54	MCE#04	32°50.491' S 138°01.555' E
55	MCE#05	32°50.516' S 138°01.505' E
56	MCE#07	32°50.503' S 138°01.429' E
57	MCE#06	32°50.511' S 138°01.439' E
58	TG#03	33°01.437' S 138°06.272' E
59	TG#04	33°01.417' S 138°06.293' E
60	TG#02	33°01.387' S 138°06.355' E
61	TG#01	33°04.386' S 138°06.360' E

Printed by Books on Demand GmbH, Norderstedt / Germany